PTEROSAURS

FLIGHT IN THE AGE OF DINOSAURS

AMERICAN MUSEUM Ö NATURAL HISTORY

NEW YORK

CONTENTS

FLIGHT IN THE AGE OF DINOSAURS

They flew with their fingers. They walked on their wings.

Some were gigantic; others could fit in the palm of a hand.

Millions of years ago, the skies were ruled by pterosaurs,

the first animals with backbones to fly under their own

power—not just leaping or gliding, but flapping their wings

to generate lift and travel through the air. Pterosaurs

(from the Greek words *pteron*, meaning "wing," and *sauros*,

"lizard") were closely related to dinosaurs, but had

streamlined bodies, narrow jaws, and long forelimbs

adapted for life in the air. Fossils reveal the vast diversity

of these extraordinary flying reptiles. More than 150

species have been discovered in excavations around the

globe. Scientists believe there are many more pterosaur

species that have yet to be discovered.

A TIMELINE OF PTEROSAURS

Pterosaurs flourished during the Mesozoic Era, the chapter of Earth's history when dinosaurs dominated the land.

TRIASSIC PERIOD
252–201 MILLION YEARS AGO

The oldest known pterosaur fossils are about 220 million years old. The earliest pterosaurs were relatively small, with long tails, short necks, and jaws lined with teeth.

JURASSIC PERIOD
201–145 MILLION YEARS AGO

At this time, a new group of pterosaurs emerged, with shorter tails and longer hand and neck bones.

Pterodactylus antiquus

Eudimorphodon ranzii

Dimorphodon macronyx

Sordes pilosus

Campylognathoides liasicus

Wukongopterus lii

Raeticodactylus filisurensis

Preondactylus buffarinii

Sauropod dinosaurs roamed the land

Rhamphorhynch muensteri

First mammal

First dinosaurs

Marine reptiles, including ichthyosaurs, dominated seas

▶ **240 MILLION YEARS AGO** ▶ 210 ▶▶▶ **200** ▶▶▶ **190** ▶▶▶ **180** ▶▶ 170 ▶▶▶ 160 ▶▶▶ 150 ▶▶

CRETACEOUS PERIOD
145–66 MILLION YEARS AGO

A wide variety of pterosaurs evolved during the Cretaceous, including the largest pterosaurs known. Many species had long, slender skulls and very long necks. Some had no teeth and huge crests.

CENOZOIC ERA
66 MILLION YEARS AGO TO PRESENT

After pterosaurs and large dinosaurs went extinct, the Age of Mammals began.

K-T EXTINCTION EVENT
Almost all large vertebrates—including pterosaurs and non-avian dinosaurs—went extinct 66 million years ago.

After pterosaurs and large dinosaurs went extinct, the Age of Mammals began.

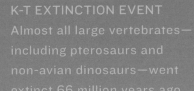

Pteranodon longiceps

Thalassodromeus sethi

Istiodactylus latidens

Anhanguera blittersdorffi

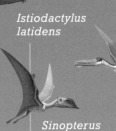

Sinopterus dongi

Jeholopterus ningchengensis

Pterodaustro guinazui

Nyctosaurus gracilis

Quetzalcoatlus northropi

▶▶ **130** ▶▶▶ **120** ▶▶▶ **110** ▶▶▶ **100** ▶▶▶ **90** ▶▶▶ **80** ▶▶▶ **70** ▶▶▶ **60** ▶▶▶ **50** ▶▶▶▶▶▶

A WIDE RANGE OF WINGSPANS

Pterosaurs all had the same basic body plan but varied dramatically in size. Some were small enough to fit in the palm of your hand, while others were as large as hang gliders. The wingspan of *Quetzalcoatlus northropi* was about 33 feet (10 m), some 40 times as long as that of the tiny pterosaur *Nemicolopterus crypticus*. The model skeleton below of a *Nemicolopterus crypticus* is shown at actual size.

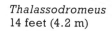

Thalassodromeus
14 feet (4.2 m)

Rhamphorhynchus
up to 6 feet (1.8 m)

Nemicolopterus
10 inches (25 cm)

Preondactylus
18 inches (45 cm)

Sordes
2 feet (60 cm)

Jeholopterus
3 feet (1 m)

Nemicolopterus crypticus
nem·ee·kol·OP·ter·us KRIP·tik·us

	WHEN	WHERE	FOOD	WINGSPAN
Nemicolopterus crypticus means "hidden flying forest-dweller."	Around 120 million years ago	Forest in what is now northeastern China	Probably insects	10 inches (25 cm)

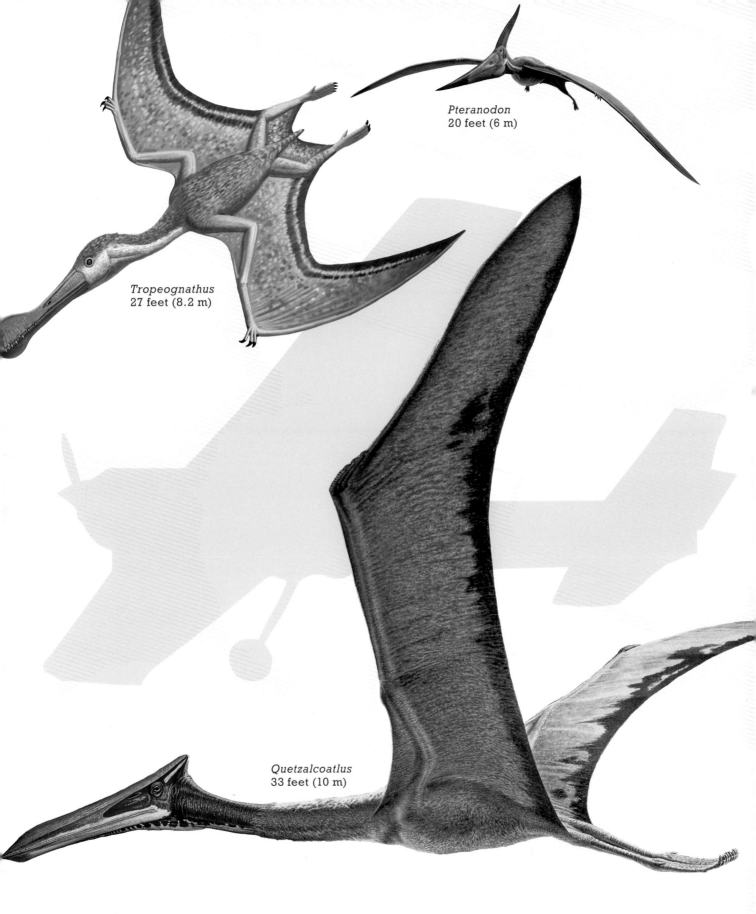

Pteranodon
20 feet (6 m)

Tropeognathus
27 feet (8.2 m)

Quetzalcoatlus
33 feet (10 m)

9

FRAGILE FOSSILS

Around 66 million years ago, at the same time that *Tyrannosaurus rex* and other large dinosaurs became extinct, pterosaurs also died out. Pterosaurs left no descendants—only fossils. But not very many fossils, particularly compared to their dinosaur cousins. Pterosaurs had especially fragile bones that preserved poorly, so pterosaur fossils are frequently incomplete. To form a picture of a particular species, paleontologists must often gather information from several fossils, or draw conclusions from related pterosaurs that are better known.

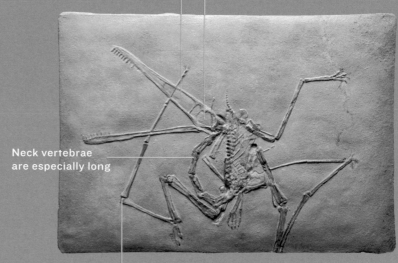

Large nostril and eye socket helped keep the skull light

Neck vertebrae are especially long

Long fourth finger supported the wing

Pterodactylus antiquus fossil

FIRST FOSSIL

Pterodactylus antiquus was the first pterosaur to be studied and described. In the late 1700s, a fossil of this unusual animal was acquired by a German ruler who kept a *Wunderkammer*, or cabinet of natural wonders, like many aristocrats of his day. The palace naturalist, Italian scholar Cosimo Alessandro Collini, examined the fossil in 1784. Collini was mystified. He was impressed by the animal's long, flexible forelimbs but could not imagine what they were for.

In the early 1800s, scientists struggled to make sense of the creature represented by the fossil above.

1800

Some naturalists thought pterosaurs might still be alive, perhaps in the shape of this animal that reportedly lived in China.

1830

One scholar thought pterosaurs lived in the sea and had long flippers like a sea turtle, a dolphin-like head, and a neck like a duck's.

1843

This picture shows a pterosaur as a flying marsupial similar to a bat.

THE FIRST "PTERODACTYL"

Scientists have puzzled over pterosaurs since the 1700s, when an unusual fossil found its way into the collection of a German prince. The first scientist to correctly identify this mysterious creature as a flying reptile was French zoologist Georges Cuvier, in 1809. Cuvier gave it a name: *ptéro-dactyle*, meaning "wing finger."

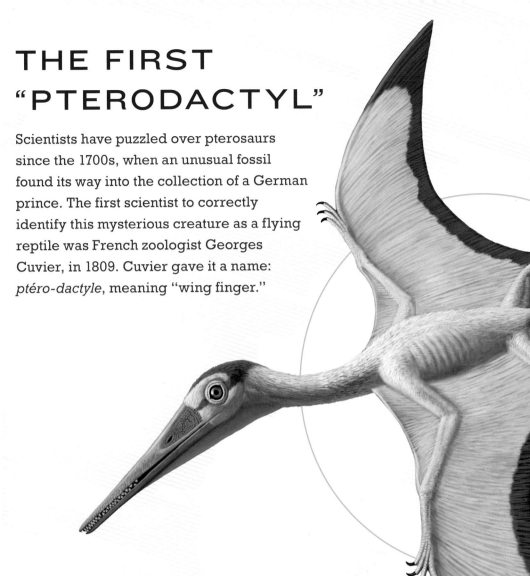

EARLY EXPERT

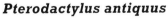

The French naturalist Georges Cuvier (1769–1832) understood early on that pterosaurs were flying reptiles. Cuvier spent years comparing the skeletons of animals, both living and extinct. In addition to pterosaurs, he identified the giant sloth *Megatherium* in South America, and the mastodon in the United States.

Pterodactylus antiquus
tair·o·DAK·til·us an·TEEK·wus

Pterodactylus antiquus caught its prey between long jaws lined with small conical teeth.

WHEN	WHERE	FOOD	WINGSPAN
Around 150 million years ago	An archipelago in what is now southern Germany	Insects or fish	Up to 5 feet (1.5 m)

A NEW DISCOVERY

While scientists studied the first "pterodactyls" in Germany, collecting fossils was becoming a popular pastime in many parts of the world. One of the most avid collectors in England was a carpenter's daughter named Mary Anning, who supported her family by finding and selling fossils near her seaside home.

In 1828, Anning discovered the remains of a new kind of pterosaur, proving that flying reptiles were varied and had a wide range. *Dimorphodon macronyx* was more robust than its German cousins, with shorter wings, a larger head, and a longer tail.

UNITED KINGDOM

Lyme Regis

THE PRINCESS OF PALEONTOLOGY

The English fossil hunter Mary Anning rarely left her seaside home at Lyme Regis in southern England. She had little formal education, and very little means. Yet by collecting and studying fossils, she earned a reputation so widespread that a German naturalist called her the princess of paleontology. At age 13, Anning unearthed the skeleton of a giant marine reptile: one of the first ichthyosaurs. Many more discoveries are based on fossils found by Anning, including Jurassic fish, plesiosaurs, and *Dimorphodon macronyx*, the first English pterosaur.

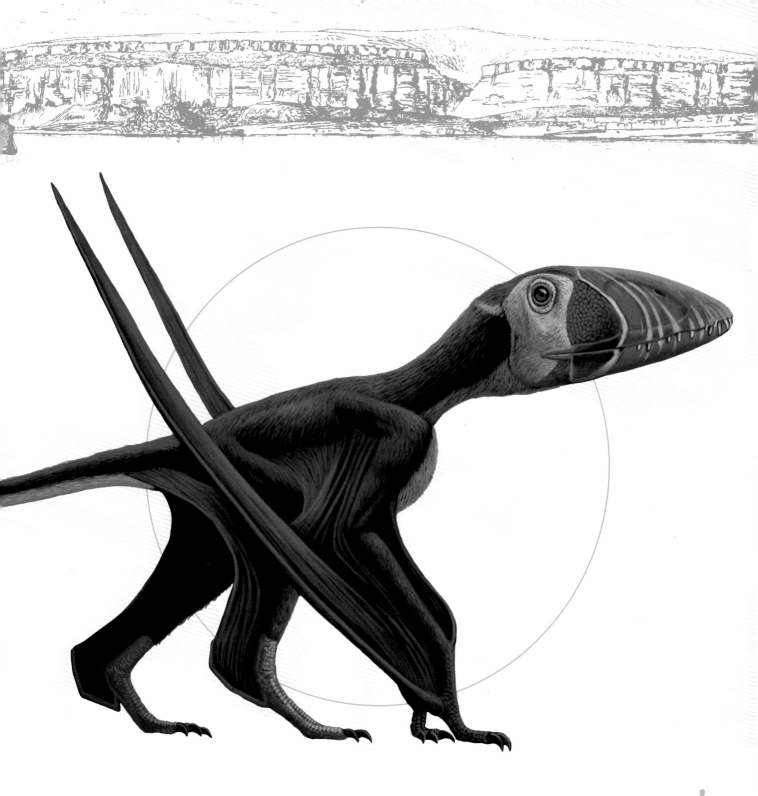

Dimorphodon macronyx
dye·MORF·o·don ma·KRON·ix

The floor of the sea where *Dimorphodon* lived was high in iron, so its fossils are colored iron-gray.

WHEN
Around 200 million years ago

WHERE

On a coast in what is now southern England

FOOD

Insects, fish, and other small vertebrates

WINGSPAN

Up to 4 feet 7 inches (1.4 m)

FINE SPECIMEN

Around 150 million years ago, a young pterosaur died, and its body sank to the bottom of a lagoon. Before the corpse could decay, layers of sediment settled on top, pressing the pterosaur flat, like a flower pressed between pages of a book. Minerals replaced the bones, so the skeleton turned to stone. For eons, this pterosaur lay hidden in a bed of limestone near Solnhofen, Germany— until a worker cut the stone from a quarry and cracked the layers apart, revealing the skeleton, along with a ghostly impression in the facing rock.

Pterodactylus antiquus fossil discovered in a bed of limestone near Solnhofen, Germany.

A quarry near Solnhofen, Germany, in the 1700s

14

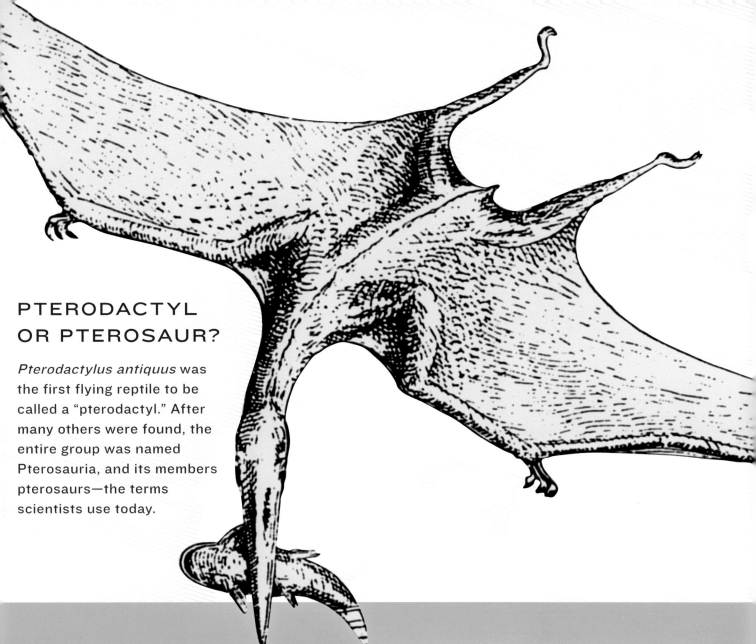

PTERODACTYL OR PTEROSAUR?

Pterodactylus antiquus was the first flying reptile to be called a "pterodactyl." After many others were found, the entire group was named Pterosauria, and its members pterosaurs—the terms scientists use today.

BURIED TREASURE

The first pterosaurs were discovered in the Solnhofen quarries of southern Germany, well known for their magnificent fossils. For centuries, the limestone of this area has been cut for paving, shingling, and lithography, a method of printing that uses stone plates. The same fine-grained stone that produced beautiful printed pictures also preserved ancient species in exquisite detail.

Solnhofen shale

Turtle

IMPERFECT EVIDENCE

Of all the pterosaurs that ever lived, only a minuscule fraction died under the right conditions to be captured as fossils. Even fewer are preserved intact. Pterosaur bones were thin and fragile, much like bird bones, and they often drifted apart, shattered, or became scrambled before they could be preserved. Pterosaur fossils are also easily damaged when extracted, transported, or prepared for study or display. Evidence is scarce, and each fossil offers just a hint of the wide array of pterosaurs that once populated the globe.

Fossil of *Preondactylus buffarinii*

WHAT COLOR WERE PTEROSAURS?

Fossils don't readily show the color of pterosaurs. No one knows what color they were, so the pterosaur colors shown in this book are borrowed from living animals with similar ways of life.

Pterosaur crests, like some bird crests, may have been brightly colored, helping members of the same species recognize each other.

Pterosaurs that lived near the shore may have been colored like sea birds, in shades of white, gray, and black.

Pterosaurs that flew through the forest may have worn tones of brown or green.

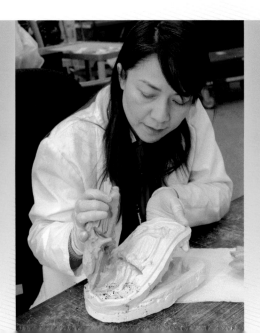

FOSSIL CASTS

Many of the pterosaur remains displayed in museums are casts: reproductions made by taking an impression of the original fossil. A well-made cast can look very much like the original, and scientists often share information by making and exchanging fossil casts. A cast can also be a useful tool. The fossil skeleton of *Preondactylus buffarinii* was lost, but scientists made a silicone rubber cast of the traces left in the rock to reconstruct the shape of the bones.

Preondactylus buffarinii
pree·on·DAK·til·us buf·a·REEN·ee·eye

Like many early pterosaurs,
Preondactylus buffarinii
had a very long tail.

WHEN	WHERE	FOOD	WINGSPAN
Around 220 million years ago	Near an inland sea in what is now northern Italy	Insects or fish	18 inches (45 cm)

Pigeon

17

PTEROSAUR EVOLUTION

Pterosaurs flourished in every period of the Mesozoic Era, the chapter of Earth's ancient history when dinosaurs dominated the land. More than 150 pterosaur species emerged and died out during that time. The earliest pterosaurs—relatively small flying reptiles with sturdy bodies and long tails—evolved into a broad variety of species. Some had long, slender jaws, elaborate head crests, or specialized teeth, and some were extraordinarily large. This family tree shows some of the most important species and how they are related.

A wide variety of pterosaurs evolved during the Cretaceous Period (145–66 million years ago), including the largest pterosaurs known. Many species had long, slender skulls with large, bony crests, toothless jaws, or very long necks.

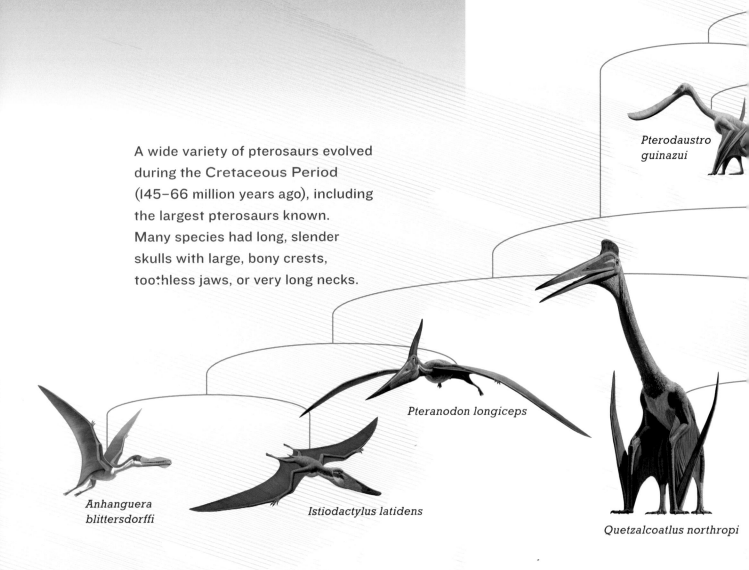

Pterodaustro guinazui

Pteranodon longiceps

Anhanguera blittersdorffi

Istiodactylus latidens

Quetzalcoatlus northropi

The earliest pterosaurs evolved during the Triassic Period (252–201 million years ago). They were relatively small and robust, with long tails, short necks, and jaws lined with teeth.

Preondactylus buffarinii

Jeholopterus ningchengensis

Campylognathoides liasicus

Eudimorphodon ranzii

Rhamphorhynchus muensteri

Wukongopterus lii

Pterodactylus antiquus

In the Jurassic Period (201–145 million years ago), a new group of pterosaurs emerged, with much shorter tails and longer necks.

Nyctosaurus gracilis

Dsungaripterus weii

Sinopterus dongi

Thalassodromeus sethi

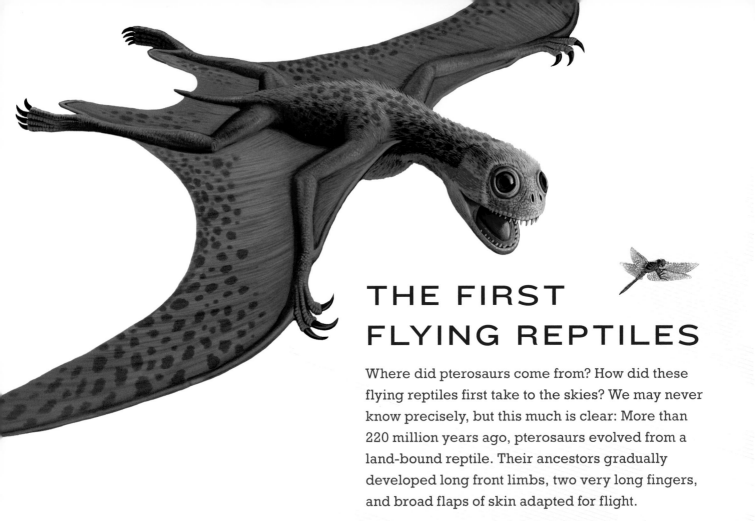

THE FIRST FLYING REPTILES

Where did pterosaurs come from? How did these flying reptiles first take to the skies? We may never know precisely, but this much is clear: More than 220 million years ago, pterosaurs evolved from a land-bound reptile. Their ancestors gradually developed long front limbs, two very long fingers, and broad flaps of skin adapted for flight.

COMPETING THEORIES

How did the first flying reptiles evolve? Based on the anatomy of pterosaurs and other gliding and flying animals, scientists have proposed several theories to explain this remarkable change.

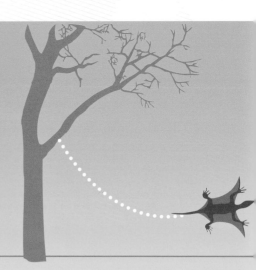

RUNNING and LEAPING?

Pterosaurs may have evolved from a reptile that ran around on its hind legs and generated lift by flapping its arms.

JUMPING from TREES?

A tree-dwelling reptile may have jumped to the ground to catch prey or flee predators. Flaps of skin helped it glide farther and later evolved into flapping wings.

FROM ARMS INTO WINGS

Through natural selection, the ancestors of pterosaurs developed traits that were helpful for gliding. Gradually these traits evolved into wings.

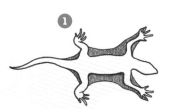

1 Skin flaps evolved on the sides of the body.

2 The fourth finger became longer, making it easier to steer.

3 Hand and finger bones, and a newly evolved bone called the pteroid, supported a flapping wing.

4 Wings became longer, and a membrane evolved on the tail for balance and steering.

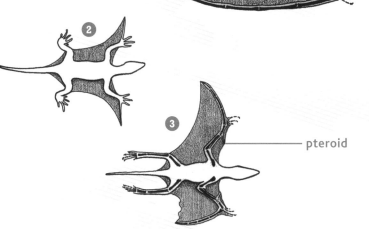

pteroid

LEAPING from TREE to TREE?

Pterosaurs may have descended from a reptile that leaped from branch to branch in the woods. Leaping evolved into gliding, and eventually into flight.

FAMILY TREE

Scientists have long debated where pterosaurs fit on the evolutionary tree. The leading theory today is that pterosaurs, dinosaurs, and crocodiles are closely related and belong to a group known as archosaurs. One of the closest early cousins of pterosaurs was probably a small terrestrial reptile known as *Scleromochlus taylori*.

FLIGHTLESS COUSIN?

The ancestor of pterosaurs may have looked something like *Scleromochlus taylori*, an ancient land-dwelling reptile around the size of a blackbird.

Scleromochlus taylori

sklair·o·MOCK·lus TAY·ler·eye

Scleromochlus taylori was a close relative of pterosaurs, with long legs, short arms, and no wings.

WHEN	WHERE	FOOD	WINGSPAN
Around 230 million years ago	Desert sand dunes in what is now northern Scotland	Perhaps insects	About 7 inches (18 cm) from nose to tail tip

Pigeon

RELATED TO DINOSAURS

Pterosaurs have many features in common with dinosaurs but sit on their own branch of the family tree.

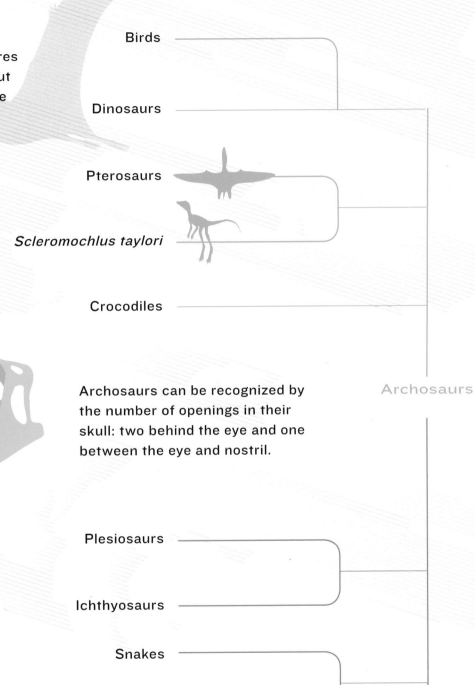

Birds

Dinosaurs

Pterosaurs

Scleromochlus taylori

Crocodiles

Archosaurs

Archosaurs can be recognized by the number of openings in their skull: two behind the eye and one between the eye and nostril.

Eye

Nostril

Plesiosaurs

Ichthyosaurs

Snakes

Lizards

INSIDE THE NEST

Were pterosaurs good parents? What did the youngest pterosaurs look like, and how did they change as they grew? For many years, the family life of pterosaurs was shrouded in mystery. Fossils of very young pterosaurs are exceptionally rare, and for good reason: Their delicate bones dissolved easily, and only by the slimmest chance were they ever carried to environments where fossils could form. Recently, however, scientists have made some groundbreaking discoveries. These remarkable fossils are beginning to give us a clearer picture of a pterosaur's first few days.

FRESH START

Like birds and most other reptiles, female pterosaurs laid eggs. Inside the egg's protective shell, the young pterosaur's skeleton developed, gradually becoming more solid and more complete.

By the time it hatched, a pterosaur had fully formed wings, and it could probably fly within a short time. In fact, scientists now think young pterosaurs survived on their own from the start, with no help from parents.

The canary-size creature at right is a very young pterosaur. Its delicate body was preserved in fine sediment, so the skeleton appears very clearly, including its minuscule fingers and toes. If this pterosaur had lived longer, it might have grown as large as a seagull.

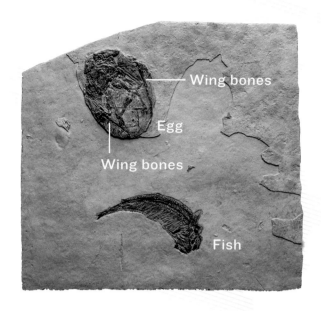

Wing bones

Egg

Wing bones

Fish

PTEROSAUR EGG

In 2004, scientists in China discovered the first fossil of a pterosaur egg, caught in a slab of shale near the remains of an ancient fish. Inside the egg, the pterosaur lay curled up, with its wings wrapped around its body. The skeleton was nearly complete, meaning the young pterosaur was almost ready to come out of its shell. Its wing bones were long and fairly solid, so soon after hatching, it would probably have been able to take off and fly.

HARD EGGS OR SOFT?

Living animals lay many different types of eggs. Birds and some turtles lay eggs with hard shells, while snakes and lizards lay soft, leathery eggs.

Prehistoric eggs varied too. The dinosaurs of the past laid hard-shelled eggs. This kind preserves more easily, so fossil eggs of dinosaurs are fairly common. Pterosaur eggs were soft-shelled, and only a few have been found so far.

CORN SNAKE
Soft egg

BLACK SWAN
Hard egg

25

LOCOMOTION ON LAND

Like other flying animals, pterosaurs of all ages spent part of their lives on the ground. What did they look like when resting or walking? Did they stand up on two feet, or crouch down on all fours? Since the time when pterosaurs were first discovered, scientific ideas about how pterosaurs stood and moved on land have changed many times. Today, almost all paleontologists agree that pterosaurs were quadrupedal—they walked on all four limbs—based on evidence from fossil tracks.

Dsungaripterus weii

PTEROSAUR AND DINOSAUR TRACKS

At a mineral mine in Morocco, scientists found layers of sandstone pitted with prehistoric tracks. In one remarkable fossil, pterosaur tracks appear with the track of a theropod dinosaur. Pterosaur footprints alternate with handprints, showing that pterosaurs got around on all fours — unlike the dinosaur, which walked on its hind legs.

WHICH WAY DID THEY WALK?

Before large numbers of pterosaur tracks were found in the 1990s, scientists trying to understand how these animals moved on land could only rely on comparisons with living animals. Today most experts agree that pterosaurs walked on all four limbs. Perhaps they could also climb vertically, as bats do. They did not hang upside down or walk upright.

1836: Scientists pictured pterosaurs crawling upward, like a bat.

1924: One expert thought pterosaurs walked upside down, like a sloth.

Pterosaur hand

Pterosaur foot

Pterosaur
hands

Pterosaur foot

Theropod foot

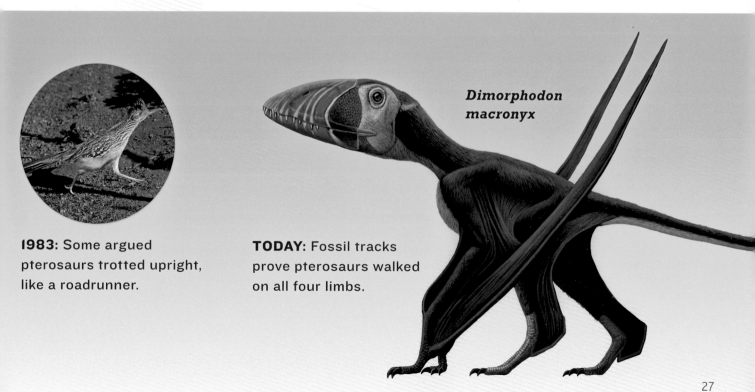

1983: Some argued pterosaurs trotted upright, like a roadrunner.

TODAY: Fossil tracks prove pterosaurs walked on all four limbs.

Dimorphodon macronyx

INTO THE AIR

For more than 100 million years in the history of life on Earth, the only animals that truly flew—not just glided—were insects. Then, around the time that dinosaurs arose, the first pterosaurs appeared in the sky. A new era began: the age of flying reptiles. Flight allowed pterosaurs to travel long distances, exploit new habitats, escape predators, and swoop down from above to seize their prey. No longer tethered to the ground, they spread across the world and branched out into an enormous array of species, including the largest animals ever to take wing.

MODEL FLYER

How well did pterosaurs fly? In the 1940s, German biologist Erich von Holst tried to find out by building a mechanical model of *Rhamphorhynchus*. Made of balsawood, wire, and tissue paper, with a rubber-band motor for flapping the wings, this model pterosaur made its first public appearance at a paleontology conference, flying around a crowded lecture hall. Von Holst compared its flight to that of a large tern. "It is assured and elegant, like a bird," he wrote, "and not at all reminiscent of bats."

Rhamphorhynchus muensteri
ram·fo·RIN·kus MOON·ster·eye

Rhamphorhynchus muensteri is one of the most widely studied pterosaurs and is represented by more than 100 fossils.

WHEN	WHERE	FOOD	WINGSPAN
Around 150 million years ago	An archipelago in what is now southern Germany	Fish	Up to 6 feet (1.8 m)

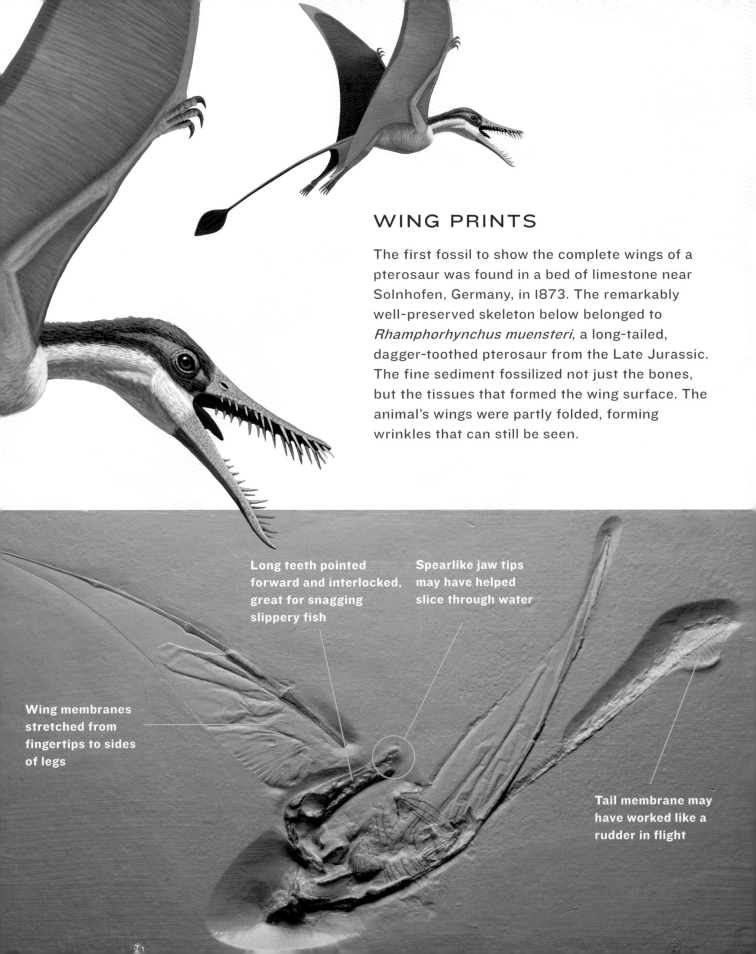

WING PRINTS

The first fossil to show the complete wings of a pterosaur was found in a bed of limestone near Solnhofen, Germany, in 1873. The remarkably well-preserved skeleton below belonged to *Rhamphorhynchus muensteri*, a long-tailed, dagger-toothed pterosaur from the Late Jurassic. The fine sediment fossilized not just the bones, but the tissues that formed the wing surface. The animal's wings were partly folded, forming wrinkles that can still be seen.

Long teeth pointed forward and interlocked, great for snagging slippery fish

Spearlike jaw tips may have helped slice through water

Wing membranes stretched from fingertips to sides of legs

Tail membrane may have worked like a rudder in flight

WING BONES

Although many animals can glide through the air, pterosaurs, birds, and bats are the only vertebrates that have evolved to fly by flapping their wings. All three groups descended from animals that lived on the ground, and their wings evolved in a similar way: Their forelimbs gradually became long, bladelike, and aerodynamic. While they have much in common, pterosaurs, birds, and bats developed the ability to fly independently. Their wings evolved along different paths, and the difference can be seen in their structure.

Istiodactylus

NOT QUITE FLYING

Pterosaurs, birds, and bats are the only vertebrates that have achieved powered flight—the ability to stay in the air through their own muscle movements. Many other vertebrates travel through air but gradually lose altitude. They are gliders, rather than flyers, in spite of their names.

Flying squirrel

Flying frog

Flying fish

Flying lemur

Flying lizard

THREE KINDS OF WINGS

The wing bones of pterosaurs, birds, and bats correspond to the bones in our arms and hands. In each of these flying animals, the bones evolved in a different way to build a wing.

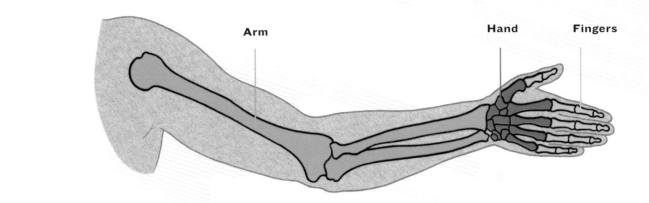

Arm **Hand** **Fingers**

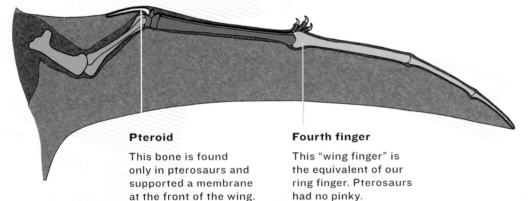

PTEROSAUR

The hand bones are long, and the fourth finger is extremely long. The wing surface is a membrane stiffened with several layers of internal fibers.

Pteroid

This bone is found only in pterosaurs and supported a membrane at the front of the wing.

Fourth finger

This "wing finger" is the equivalent of our ring finger. Pterosaurs had no pinky.

BAT

The hand bones and four finger bones are very long. They support the wing surface, made of a thin, flexible membrane.

BIRD

The finger bones are reduced and fused. The wing surface is made of feathers.

BUILT TO FLY

Imagine *Pteranodon*, an immense pterosaur with a 20 foot (6 m) wingspan, soaring through the skies above an ancient seaway. How could such a large creature fly through the air? Like other flying animals, pterosaurs generated lift with their wings. They needed to perform motions similar to those of birds and bats, but their wings evolved independently, developing their own distinct aerodynamic structure. A combination of several adaptations allowed these animals to fly.

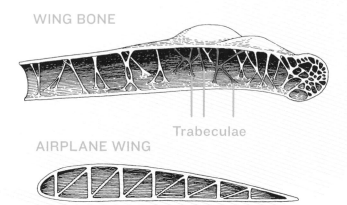

Pteranodon longiceps
ter-AN-o-don LON-ji-seps

With its long, dramatic crest, the giant *Pteranodon longiceps* is one of the most recognizable pterosaurs.

WHEN	WHERE	FOOD	WINGSPAN
Around 85 million years ago	Over a large seaway covering what is now central North America	Fish	Up to 20 feet (6 m)

HOLLOW BONES

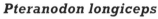

WING BONE

Trabeculae

AIRPLANE WING

Large pterosaurs needed strong limbs to get off the ground, but thick bones would have made them too heavy. The solution? A pterosaur's wing bones were hollow tubes, with walls no thicker than a playing card. Like bird bones, they were flexible and lightweight, while strengthened by internal struts. A network of struts called trabeculae inside the bones of pterosaurs and birds helped make them stronger, as in an airplane wing.

WING SHAPE

Perhaps the most important adaptation that allowed pterosaurs to fly was the shape of their wings. Pterosaur wings are thicker in front—the part that faces the direction of travel. It's how air flows around this shape—creating a pressure difference between the top and bottom of the wing—that generates lift.

Wukongopterus lii

WING FLAPPING

How often would pterosaurs have had to flap their wings? In part, that depends on their size. Large pterosaurs, like large birds today, would tire out after a few minutes of flapping because they encountered more air resistance, or drag. Long, narrow wings are an adaptation that reduces drag. Lighter-weight pterosaurs could afford stouter wings. They would have flapped more frequently but they had much greater control over their speed and direction.

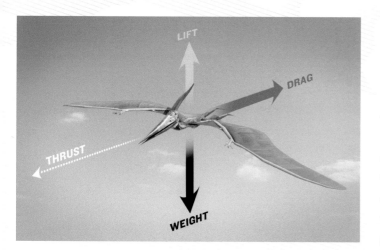

INSIDE
THE WINGS

Recent discoveries show that pterosaur wing membranes were more than simple flaps of skin. Long fibers extending from the front to the back of the wings formed a series of stabilizing supports, so the membranes could be stretched taut, or folded up like a fan. Separate muscle fibers helped pterosaurs adjust the tension and shape of their wings, and veins and arteries kept the wings nourished with blood.

WING FIBERS

MUSCLE

BLOOD VESSELS

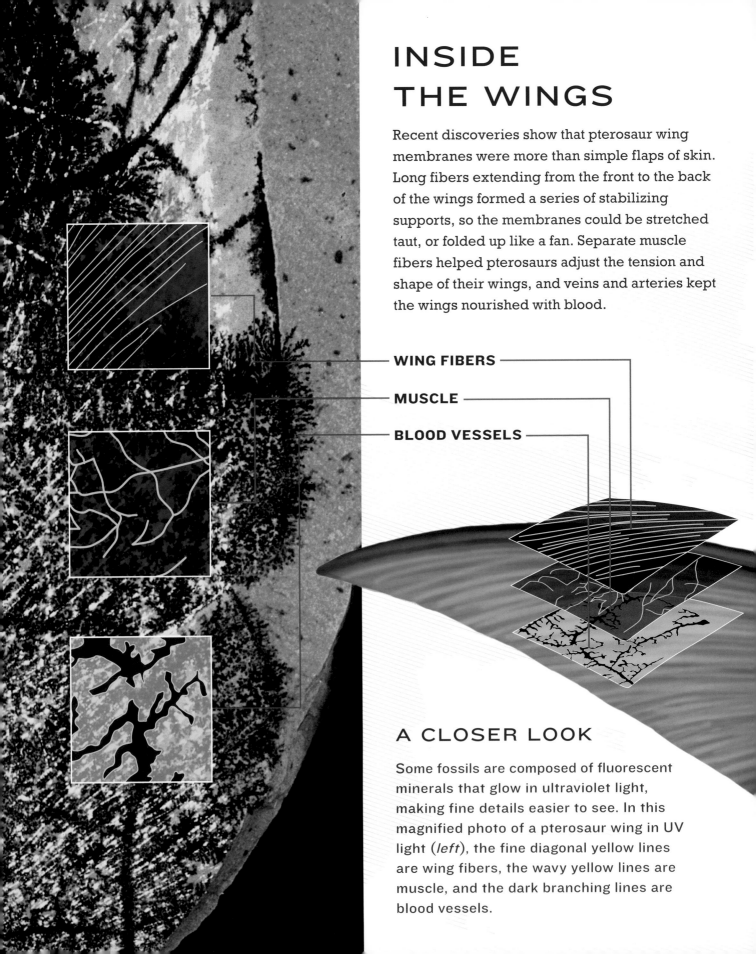

A CLOSER LOOK

Some fossils are composed of fluorescent minerals that glow in ultraviolet light, making fine details easier to see. In this magnified photo of a pterosaur wing in UV light (*left*), the fine diagonal yellow lines are wing fibers, the wavy yellow lines are muscle, and the dark branching lines are blood vessels.

WELL SUPPORTED

How did pterosaur wings keep their shape? A flight membrane without support could sag and flutter out of control when not extended fully. But like the wings of birds and bats, pterosaur wings were engineered for stability.

BIRD

Rows of feathers form a firm yet flexible wing surface.

BAT

Four long fingers support the wing membrane.

PTEROSAUR

Rows of fibers reinforce the wing.

WARM AND FUZZY

While searching for fossil insects in Kazakhstan in the 1960s, a Russian paleontologist discovered a new kind of pterosaur. It had tapering jaws, short, sharp teeth, and fairly short wings, like many of its relatives. But the fine sediment that buried this pterosaur also preserved a more striking detail: traces of fibers that looked like fur. The discovery of the species *Sordes pilosus*, or "hairy devil," proved that pterosaurs had a fuzzy coat and were probably warm-blooded.

Sordes pilosus SOR·deez pi·LO·sus	WHEN	WHERE	FOOD	WINGSPAN
Fossils of *Sordes pilosus* proved that pterosaurs' bodies were covered with hairlike fibers instead of scales.	Around 160 million years ago	Near a lake in what is now southern Kazakhstan	Small animals such as fish	2 feet (60 cm)

WELL INSULATED

Birds and bats have a high body temperature and high metabolism, which help provide energy for flight. Pterosaurs likely did too. Unlike many other reptiles, whose body temperature changes with their environment, pterosaurs were probably endothermic—that is, their temperature was controlled from within. Their fuzzy coat provides evidence for this. Like feathers on birds and fur on mammals, it likely helped them retain heat.

Close-up of fibers

Top of skull

PLUSH PELT

The head, neck, body, and upper legs of *Sordes pilosus* were carpeted with fur-like fibers, visible in this photo of the top of a fossil skull (*left*). The strands were about as thick as a human hair. Although they look like mammal hair, they evolved independently; paleontologists call them pycnofibers.

DRAGON OF THE DAWN

Some of the largest pterosaurs ever to take to the skies arose toward the end of the era of flying reptiles, between 90 and 66 million years ago. *Dawndraco kanzai*—"Kansas dragon of the dawn"—is one of the most complete large pterosaurs discovered so far. This long-winged, short-tailed, toothless creature was a close relative of *Pteranodon longiceps*. During their heyday, *Dawndraco*, *Pteranodon*, and their oversize cousins soared over a seaway that covered the North American Great Plains.

Dawndraco kanzai
don·DRAK·o KANZ·eye

DISCOVERED IN KANSAS

Large pterosaurs like *Dawndraco* and *Pteranodon* were first discovered in western Kansas, near a chalk formation called Monument Rocks.

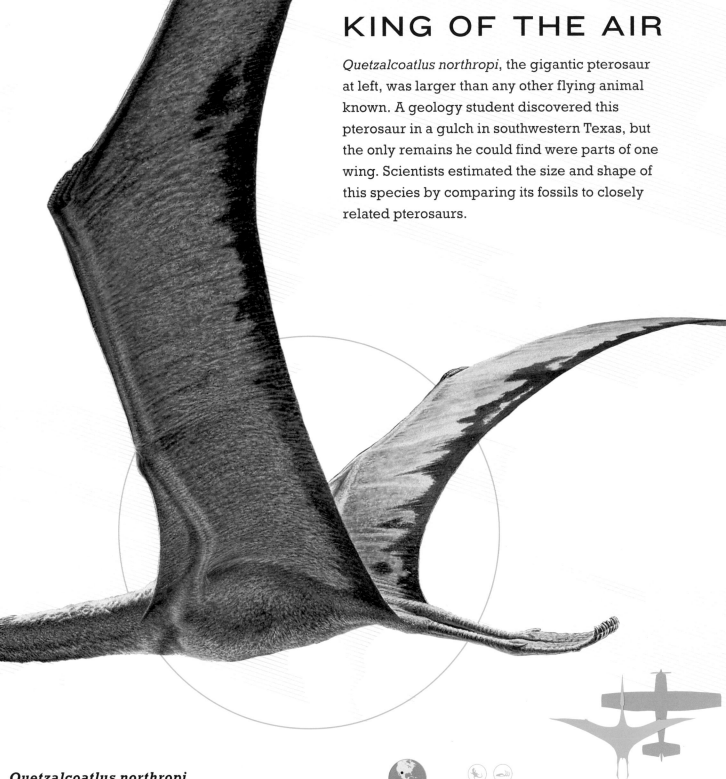

KING OF THE AIR

Quetzalcoatlus northropi, the gigantic pterosaur at left, was larger than any other flying animal known. A geology student discovered this pterosaur in a gulch in southwestern Texas, but the only remains he could find were parts of one wing. Scientists estimated the size and shape of this species by comparing its fossils to closely related pterosaurs.

Quetzalcoatlus northropi
ket·zel·KWAT·a·lus NORTH·rup·eye

Quetzalcoalus northropi was named after Quetzalcoatl, a Mexican god of the air, and industrialist Jack Northrop, known for developing a stealth aircraft called the flying wing.

WHEN	WHERE	FOOD	WINGSPAN
Around 67 million years ago	On a plain in what is now western Texas	Possibly small vertebrates or carrion	Around 33 feet (10 m)

LONG TALL PTEROSAUR

Fossils show that the wings of *Quetzalcoatlus northropi* were as long as some airplane wings, and the dimensions of related species suggest its neck and jaws were also especially long. Scientists estimate this pterosaur towered around 16 feet (5 m) when resting on all four limbs.

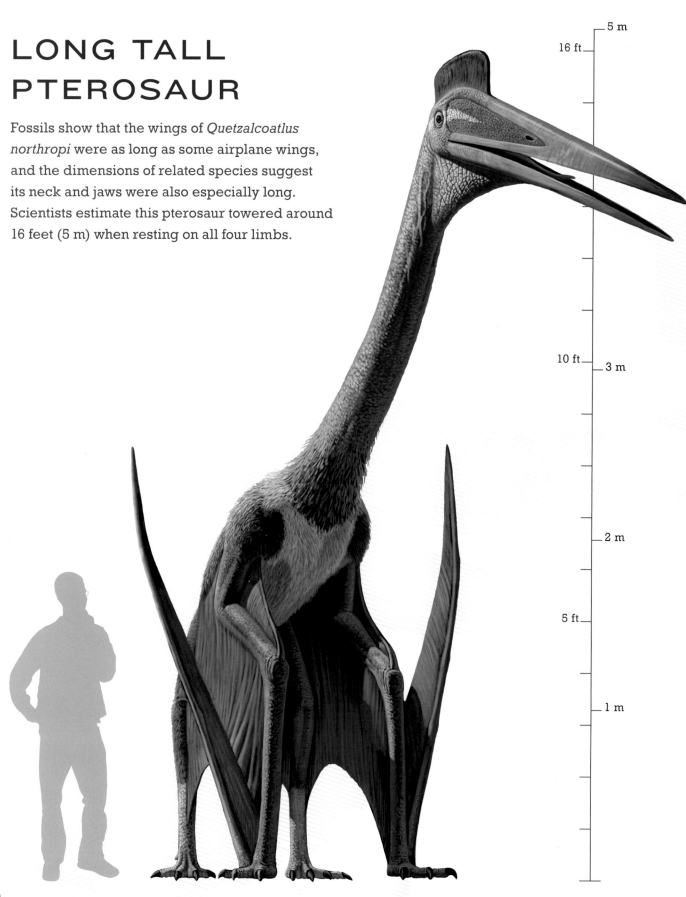

5 m

16 ft

10 ft — 3 m

2 m

5 ft

1 m

BIG DISCOVERY

While studying ancient sediments in 1971, Douglas Lawson discovered this giant pterosaur in Big Bend National Park, Texas. So far, the only known fossils of *Quetzalcoatlus northropi* are remains of one wing, though Lawson also found smaller pterosaurs with similar features about 30 miles (48 km) away.

Big Bend National Park

MANY KINDS OF CRESTS

The most dazzling feature on many pterosaurs was a spectacular head crest. These blades could stick up, back, forward, or down, depending on the species. While many pterosaurs had no crests, or very small ones, some were truly immense. What were these crests for? Scientists hotly debate their evolutionary function. Their eye-catching appearance suggests that crests evolved for display. But did they assist in identification, mating, cooling, steering, ... or something else entirely? It's impossible to say for sure.

Anhanguera

DISK-SHAPED

Some had rounded disk shapes on their snouts, on both their upper and lower jaws, like *Anhanguera blittersdorffi*.

LOW RIDGE

Some had long, low ridges running down the middle of their heads, like *Dsungaripterus weii*.

Dsungaripterus

Biggest Crest Ever Found

Thalassodromeus sethi had a crest three times larger than the entire rest of its skull, when seen from the side. Indeed, it had the largest crest of any known vertebrate, or animal with a backbone.

Crest

Beak

Eye socket

Skull opening

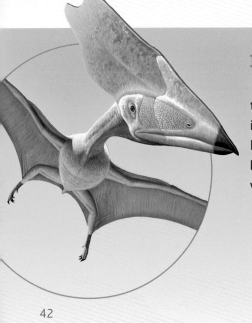

WEDGE-SHAPED

Some had prominent crests on the front of their heads, like *Raeticodactylus filisurensis*. This species is a rare example of an early pterosaur with a crest — though others might have had crests of soft tissue, which does not readily preserve.

Raeticodactylus

Pteranodon

KNIFE-SHAPED

Some had long, dagger-shaped blades jutting out of the back of their heads, like *Pteranodon longiceps*.

FAN-SHAPED

Some had giant, sail-like extensions several times bigger than the rest of their heads, like *Tupandactylus imperator*.

Tupandactylus

Who Had Them?

Crests were not confined to one single group of pterosaurs; they evolved in many groups. Among early, long-tailed species, crests were rare. But over time, pterosaur crests became larger and more flamboyant, peaking in the giant pterosaurs of the Late Cretaceous.

Eudimorphodon

TRIASSIC
252–201 million years ago

Early pterosaurs tended to have no crests, or small ones.

Germanodactylus

JURASSIC
201–145 million years ago

Later, in the Jurassic, pterosaurs with crests became more common.

Nyctosaurus

CRETACEOUS
145–66 million years ago

The biggest crests, and the biggest pterosaurs, evolved in the Late Cretaceous.

WHAT'S THE POINT?

Why did pterosaurs have crests? They must have had a specific use, or they would not have evolved repeatedly over millions of years. Scientists have many theories about why crests evolved. Two top contenders focus on the benefits of a dramatic visual appearance. But crests may well have had multiple functions.

THEORY 1

Species recognition
Pterosaurs might have used crests to identify members of their own species.

THEORY 2

Sexual selection
Pterosaurs might have used crests to attract mates.

One of us! One of us!

Flashy crests might have helped the entire species by speeding group recognition. Rapidly recognizing each other's crests could have helped pterosaurs stick together, spot enemies quickly, and find potential mates.

Examining the Evidence

If crests evolved for species identification, different species in the same region might be expected to display distinctive crests that helped them tell each other apart. Some fossil evidence supports this theory. Several crested species that lived at the same time and place in Brazil (*below*) had very different crests. Similar groupings have been found in China.

Tupuxuara

Anhanguera

Tapejara

Thalassodromeus

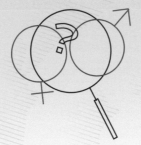

What good is looking good?

Crests might have served to attract mates. Evolution favors anything that improves one's chances with the opposite sex. If crests helped attract mates, crested individuals would leave behind lots of crested offspring, driving the evolution of crested pterosaurs. Charles Darwin called this process "sexual selection."

Examining the Evidence

According to Charles Darwin, sexual selection requires a clear difference between genders. Unfortunately, it's hard to tell from a fossil whether an animal was male or female. And male and female pterosaurs often both had crests, suggesting sexual selection was not their primary function. Sexual selection involves interactions among an entire population, so scientists continue to gather evidence.

PEACOCKS

Why do male peafowl have such incredible tails? Peahens prefer them—so their babies inherit flashy tails.

"PEACOCKING"

Lacking tail feathers, humans often use showy clothing to attract mates. Pterosaur crests might have served the same purpose.

COOLING AND STEERING

Most scientists think pterosaurs used crests either to recognize members of their own species or to attract mates. But they could have had multiple other functions, in addition to display.

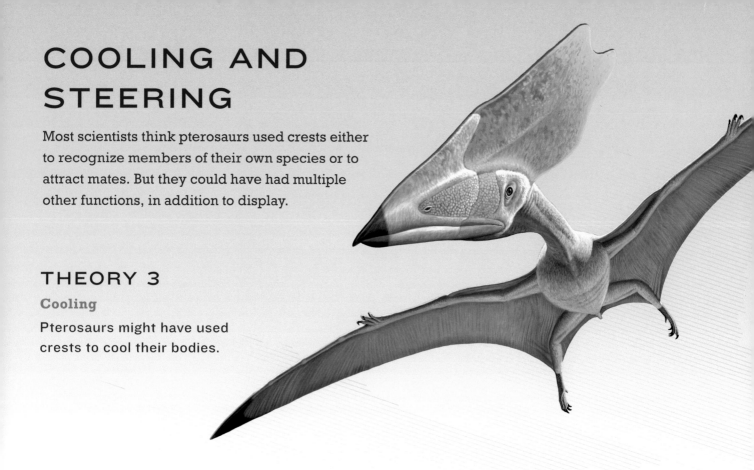

THEORY 3
Cooling
Pterosaurs might have used crests to cool their bodies.

THEORY 4
Steering
Pterosaurs might have used crests to turn or slow down.

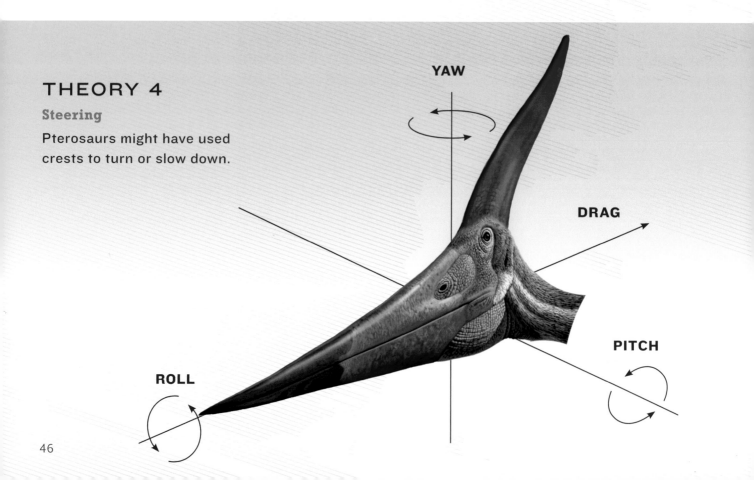

YAW

DRAG

PITCH

ROLL

Cooling with Blood

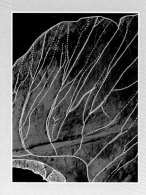

The huge, flat crest of *Thalassodromeus sethi* (*left*) contains a network of branching channels. These visible grooves suggest that, during life, a network of blood vessels covered the crest below the skin. These channels might have sent warm blood to the crest's surface, cooling the animal as it flew.

Examining the Evidence

If your body gets too hot, blood flows to your skin to release excess heat, causing you to flush red. The same thing might have happened with pterosaur crests—though blood flow to the crest could have served some other purpose.

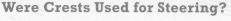

Were Crests Used for Steering?

Look at the rudder on a ship, or at an airplane's tail and wing flaps, and it's natural to wonder whether pterosaur crests had a similar function. For many years, this was a leading theory—but new evidence has raised doubts.

Examining the Evidence

In 2007, researchers put model pterosaur heads in a wind tunnel to test the aerodynamic effects of crests. They found that turning the head to block oncoming air created drag, wasted energy, put a lot of strain on the neck, and obscured vision. They concluded that crests probably did not evolve for steering, which could be done better by the wings.

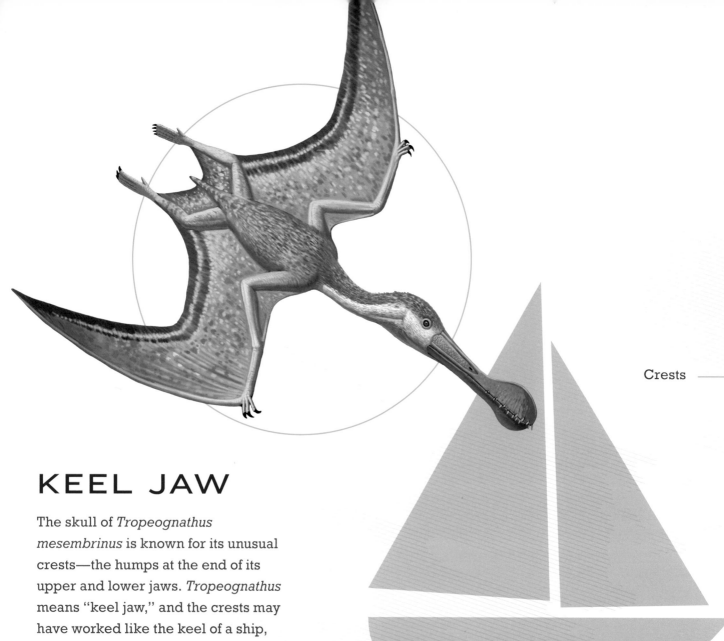

Crests

KEEL JAW

The skull of *Tropeognathus mesembrinus* is known for its unusual crests—the humps at the end of its upper and lower jaws. *Tropeognathus* means "keel jaw," and the crests may have worked like the keel of a ship, helping this pterosaur keep its balance as it plunged its snout in the water while hunting for fish.

Boat keel

Tropeognathus mesembrinus
trop·ee·og·NAY·thus mes·em·BREEN·us

The largest *Tropeognathus mesembrinus* discovered to date has a head about 3 feet (1 m) long.

WHEN
Around 110 million years ago

WHERE
Near an inland sea in what is now Brazil

FOOD
Fish

WINGSPAN
Up to 27 feet (8.2 m)

Eye socket

Teeth

FOLLOW THE TEETH

What did pterosaurs eat? With so many different kinds of pterosaurs, there is no single answer. Some species had big, sharp teeth clearly suited to stabbing prey. Others had no teeth at all and probably ate fruit. And others showed extreme modifications comparable to a wide range of animals living today, indicating highly specialized diets.

Hornbill

FRUIT EATER

Most pterosaurs ate meat, but the relatively small *Tapejara* probably ate fruit and seeds. Why do researchers think this? Its beak is similar to fruit-eating birds like parrots and hornbills.

A crest on the front of this pterosaur's head may have helped part leaves when searching for fruit, as the hornbill does today, or served for display.

Pointed beak turns down like a hornbill or parrot— the only known pterosaur jaw with this shape.

A space remained between the jaws when the tips were touching. Some have speculated that this gap might have helped it pluck and carry soft fruit.

Tapejara

A raised bump inside the beak may have helped crush seeds, cones, and hard fruit coverings.

Tapejara wellnhoferi
ta-pa-JAR-a wel-n-HOF-er-eye

Tapejara means "old being" in the Tupi language, from what is now Brazil; honors German paleontologist Peter Wellnhofer.

WHEN	WHERE	FOOD	WINGSPAN
Around 110 million years ago	Near a lagoon in what is now Brazil	Fruit, seeds, cones	About 6 feet (1.8 m)

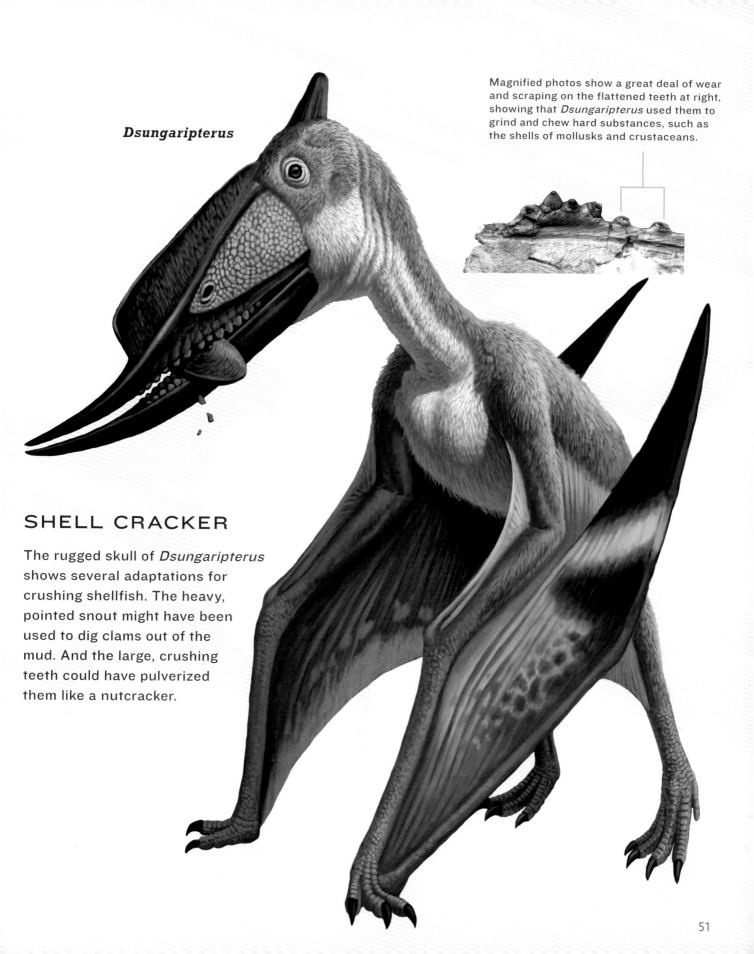

Dsungaripterus

Magnified photos show a great deal of wear and scraping on the flattened teeth at right, showing that *Dsungaripterus* used them to grind and chew hard substances, such as the shells of mollusks and crustaceans.

SHELL CRACKER

The rugged skull of *Dsungaripterus* shows several adaptations for crushing shellfish. The heavy, pointed snout might have been used to dig clams out of the mud. And the large, crushing teeth could have pulverized them like a nutcracker.

FILTER FACE

The teeth of *Pterodaustro* were so thin they resembled the bristles of a brush—and the animal had about a thousand of them, which lined the entire lower jaw. But these teeth were not for biting. Instead, the animal likely scooped up water and strained it for food. As the water flowed through its teeth, tiny animals were filtered out and swallowed. Flamingos do this today.

SIXTY TEETH PER INCH!

The filter-like teeth of *Pterodaustro* were packed so closely that a single inch of the jaw might contain 60 teeth (24 per centimeter).

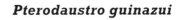

Pterodaustro guinazui

A FILTER-FEEDING WHALE

Baleen whales, which filter food from seawater, evolved highly specialized, fibery filters known as baleen. *Pterodaustro* teeth likely served the same function.

Pterodaustro guinazui tair-o-DOW-stro gee-NA-zoo-eye	WHEN	WHERE	FOOD	WINGSPAN
This pterosaur had about 1,000 teeth in its bill, which it used to feed by filtering small animals out of the water.	Around 100 million years ago	Along the shores of lakes in what is now central Argentina	Small arthropods, crustaceans, and mollusks	About 8 feet (2.5 m)

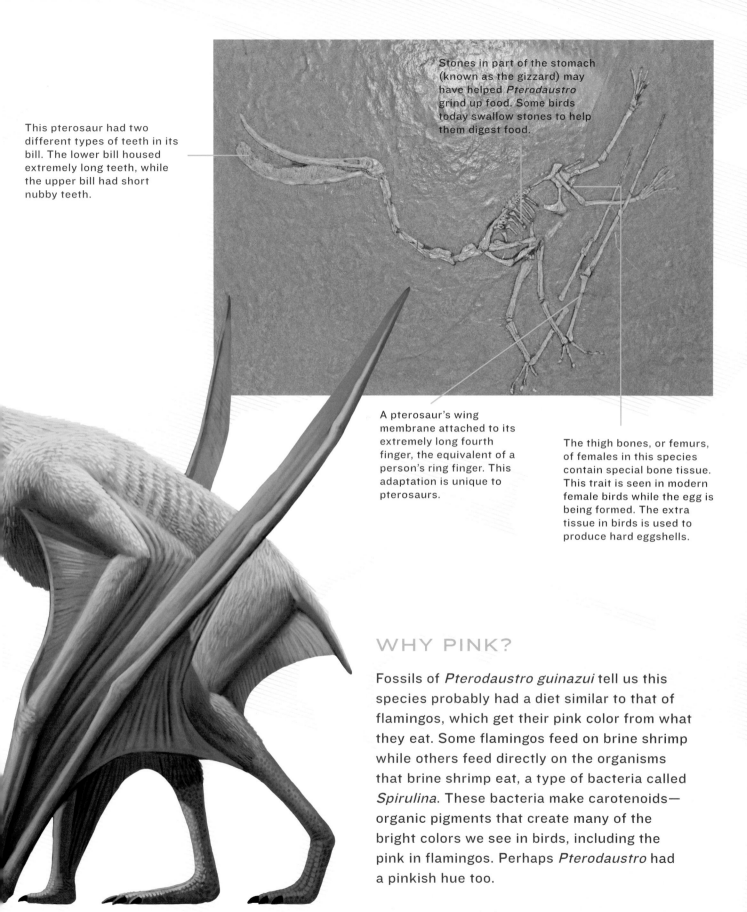

This pterosaur had two different types of teeth in its bill. The lower bill housed extremely long teeth, while the upper bill had short nubby teeth.

Stones in part of the stomach (known as the gizzard) may have helped *Pterodaustro* grind up food. Some birds today swallow stones to help them digest food.

A pterosaur's wing membrane attached to its extremely long fourth finger, the equivalent of a person's ring finger. This adaptation is unique to pterosaurs.

The thigh bones, or femurs, of females in this species contain special bone tissue. This trait is seen in modern female birds while the egg is being formed. The extra tissue in birds is used to produce hard eggshells.

WHY PINK?

Fossils of *Pterodaustro guinazui* tell us this species probably had a diet similar to that of flamingos, which get their pink color from what they eat. Some flamingos feed on brine shrimp while others feed directly on the organisms that brine shrimp eat, a type of bacteria called *Spirulina*. These bacteria make carotenoids— organic pigments that create many of the bright colors we see in birds, including the pink in flamingos. Perhaps *Pterodaustro* had a pinkish hue too.

CREATURE CATCHER

Long, sharp teeth indicate *Scaphognathus* was a predator; it might have captured small animals, including insects.

Scaphognathus crassirostris

Pteranodon longiceps

STUCK IN ITS THROAT

This rare *Pteranodon* fossil contains another fossil—the bones of a fish the pterosaur had in its mouth when it died!

HOW DID PTEROSAURS EAT?

Some chased insects. Some foraged for fruit. With over 150 known species, there is no single answer to the question of what pterosaurs ate. You would have to look at a wide variety of animals, from birds to fish to bats, to find modern-day comparisons for every pterosaur.

Jeholopterus probably chased insects like a bat.

Dsungaripterus may have crushed shellfish with its teeth like a wolffish.

Pterodaustro may have strained food from water like a flamingo.

Pteranodon probably dived into the water for fish like a pelican.

Tapejara may have eaten fruit with its curved beak like a parrot.

PTEROSAURS AND WATER

While pterosaurs ruled the skies, many also spent their days near the water. Much as shore birds do today, some pterosaurs soared above oceans and lakes, probably diving to catch fish or other sea animals. Others may have walked through shallow waters, scanning for shellfish. Thanks to those rare occurrences when fragile plants and animals are preserved as fossils, scientists can peer back into the depth of time and summon visions of a forgotten world.

Thalassodromeus sethi

CATCHING FISH

The long, thin jaws of *Thalassodromeus* seem nicely streamlined for dipping into the water to snatch fish. When the pterosaur did catch a fish, its flat crest may have helped it slice through the water without wiping out.

Thalassodromeus sethi
tha·lass·a·DRO·me·us SETH·eye

Thalassodromeus means "sea runner," referring to the animal's presumed ability to glide over the water while dipping its narrow jaws to snatch fish on the surface.

WHEN
Around 110 million years ago

WHERE
Near a lagoon in what is now Brazil

FOOD
Fish

WINGSPAN
About 14 feet (4.2 m)

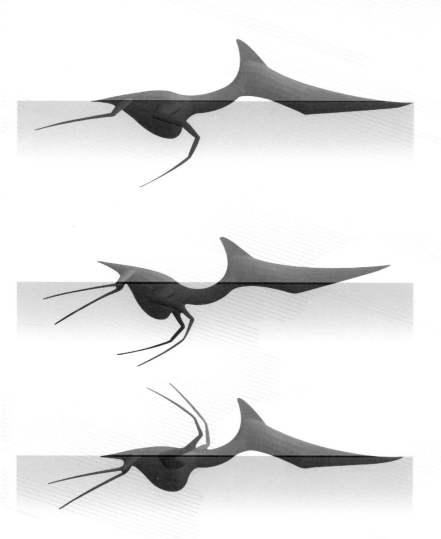

HOW DID PTEROSAURS SWIM?

Did pterosaurs paddle like ducks, with heads held high? Probably not. Most pterosaurs had big heads and short, stubby bodies. It would have been difficult even to lift their heads out of the water. And their necks were not flexible enough to fold up in an S-shape and swim like a swan.

Computer-generated models show possible positions of *Pteranodon* in water

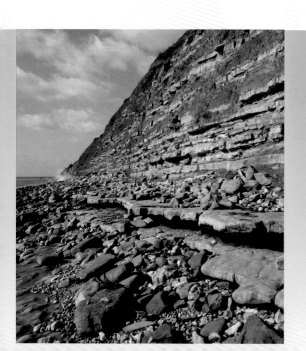

WHY WATER?

Almost all pterosaurs known today lived near water. But pterosaurs probably once lived everywhere from deserts to jungles to mountaintops, like birds today. The only ones we know about, however, are those that fossilized. And the best fossil-hunting is in limestone from former seabeds, which forms when layers of fine sediment settle to the bottom of the ocean over thousands of years—frequently preserving fossils.

A SUPERB SITE

One of the world's best sites for pterosaur fossils—the Araripe Basin in Brazil—formed as the continents of South America and Africa tore apart during the Cretaceous Era and the space between them became the South Atlantic Ocean. Rift valleys like the Araripe Basin filled with water and became home not only to pterosaurs but also to dinosaurs, crocodiles, fish, crabs, shrimp, insects, and many species of plants.Fossils found today at the Araripe Basin are often extremely well preserved. Consider the fossil shown below of *Anhanguera santanae*. When the animal died around 110 million years ago, the body was buried in fine sediment, which slowed down decomposition and allowed fossilization to begin. As the mud gradually settled, a hard shell formed around the remains and protected them. As a result, this animal was exquisitely preserved as a three-dimensional fossil.

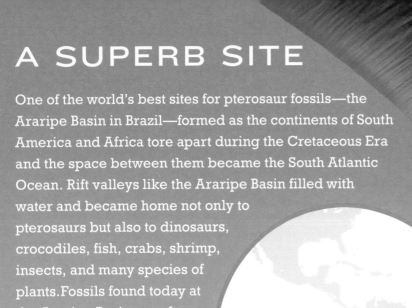

BRAZIL
Araripe Basin

Most pterosaur bones are flattened during fossilization, but the skeleton of this *Anhanguera santanae* was protected inside a hard shell that formed as mud around the animal's remains gradually turned to stone.

Eye socket

Skull is long, with prominent teeth possibly used for catching fish

Vertebrae filled with air pockets, which makes them light

Large skull opening

Anhanguera santanae
ahn·han·GWER·a san·TAN·ay

The skull of *Anhanguera santanae* was twice as long as its body.

WHEN	WHERE	FOOD	WINGSPAN
Around 110 million years ago	Near a lagoon in what is now Brazil	Fish	More than 13 feet (4 m)

UPPER LAYER: NODULES

Like many fossil sites, the Araripe Basin has several layers. The upper layer, the Romualdo Formation, features distinctive, round nodules. Here, calcium collected around dead plants and animals that sank into the soupy mud. Sometimes, a hard shell, or calcareous nodule, formed around them. These rocky coverings preserved fossils in three dimensions.

LOWER LAYER: SLABS

The Crato Formation, situated below the Romualdo Formation, formed about five million years earlier, when the Araripe Basin filled with freshwater to form a giant inland lake. Insects, plants, and pterosaurs settled to the bottom, were covered with sediment, and fossilized. Many delicate fossils were pressed between layers of fine sediments that would become limestone, like flowers in a book.

Brachyphyllum plant fossil in a nodule

Crato dragonfly fossil

HOW DO FOSSILS FORM?

Fossils can look like the bones of a living animal—but they're actually rocks that form when minerals take the place of bone or other parts of an organism. This process preserves remains for millions of years.

Nemicolopterus
crypticus

Nemicolopterus crypticus fossil

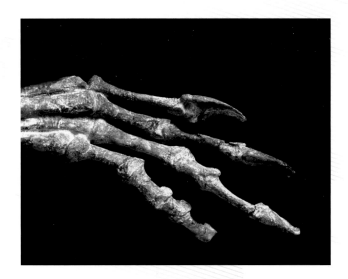

BODY FOSSILS are what we commonly think of as fossils: hardened remains of body parts like bones, shells or soft tissue. This fossil foot is from the pterosaur *Dimorphodon*.

TRACE FOSSILS are not remains of an animal itself but instead a record of its activities. Such fossils include footprints—such as these pterosaur tracks—and coprolites (fossilized droppings).

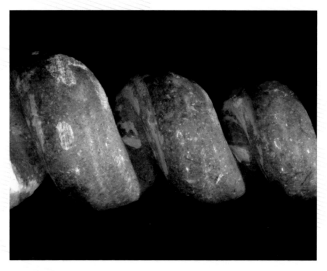

EXTERNAL MOLDS form when an organism—here, a mollusk—leaves an imprint in the sediment and that imprint is preserved.

INTERNAL MOLDS form when the inside of a shell, bone, or other animal part fills with minerals. The original hard tissue later dissolves, leaving a fossil in the shape of the space inside.

THE NEXT BIG THING?

For every pterosaur fossil that has been discovered, many more still lie buried in fossil beds around the globe. Between 2011 and 2012, paleontologists working in association with the American Museum of Natural History unearthed an exciting new pterosaur fossil in Transylvania, Romania. Fragments of several tremendous pterosaurs have been found in the same region. Scientists think a variety of giant pterosaur species lived in the area at the end of the Cretaceous Period, just before all pterosaurs became extinct. With each new discovery, our understanding of how these ancient reptiles lived will continue to change.

The neck vertebra at left measures 9 inches (23 cm) wide by 10 inches (25.5 cm) tall. The fossil above is a wristbone. Scientists estimate this pterosaur was as big as *Quetzalcoatlus* but much more massive.

Pterosaur discovery site in the Hateg Basin, Romania

Hatzegopteryx thambema

BIG AND BURLY

Only a few fossil bones have been found of one recently discovered Romanian pterosaur, and the species has not yet been named. Without more parts of the skeleton, it's difficult to tell how this animal looked in life, but it may have resembled *Hatzegopteryx thambema* (*left*), another enormous pterosaur that lived in the same area around the same time.

Romanian pterosaur	WHEN	WHERE	FOOD	WINGSPAN
A Romanian pterosaur discovered in 2012 was heavier-set than other large species, with a robust body and a wide, sturdy neck.	Around 67 million years ago	An island in what is now central Romania	Perhaps small dinosaurs and other reptiles	More than 33 feet (10 m)

This book is published by the American Museum of Natural History, Ellen V. Futter, President, in conjunction with the Museum's exhibition *Pterosaurs: Flight in the Age of Dinosaurs*, on view at the Museum from April 5, 2014 through January 4, 2015.

Pterosaurs: Flight in the Age of Dinosaurs was curated by Mark A. Norell, Curator and Chair, Division of Paleontology, American Museum of Natural History, and Alexander W. A. Kellner, Associate Professor, Museu Nacional/Universidade Federal do Rio de Janeiro, Brazil, with the assistance of Michael Habib.

American Museum of Natural History exhibitions are directed by Michael Novacek, Senior Vice President and Provost of Science, and David Harvey, Senior Vice President of Exhibition.

Written and designed by members of the Exhibition Department: Margaret Dornfeld, Lead Writer; Martin Schwabacher, Writer; Sasha Nemecek, Senior Editor; Lauri Halderman, Senior Director of Exhibition Interpretation; Dan Ownbey, Graphics Manager; Elizabeth Anderson, Kelvin Chiang, Joshua Marz, Designers; Catharine Weese, Director of Graphics; José Ramos, Graphics Research Supervisor; Melissa Posen, Senior Director of Exhibition Operations.

Published by the Global Business Development Department: Ellen Gallagher, Senior Vice President and Chief Financial Officer; Sharon Stulberg, Senior Director; Will Lach, Director.

The Museum gratefully acknowledges the **Richard and Karen LeFrak Exhibition and Education Fund**.

Generous support for *Pterosaurs: Flight in the Age of Dinosaurs* has been provided by **Mary and David Solomon**.

Book design by Atif Toor, adapted from the exhibition.

14 15 16 17 18   6 5 4 3 2 1

Printed in the United States of America.
ISBN 978-0-9852016-3-0

IMAGE CREDITS

Unless otherwise noted, all images copyright © American Museum of Natural History (AMNH). Unless otherwise noted, all pterosaur illustrations are drawn by Raúl Martín.

Pages 8–9 Big and Small. *Nemicolopterus crypticus* model: © AMNH; *Quetzalcoa? northopii:* © John Sibbick. **Pages 10–11 Fragile Fossils.** *Pterodactylus antiquus* fos AMNH FR 5134; 1800 pterosaur: Bibliothèque Centrale MNHN Paris; 1830 pterosaur AM Library; 1843 pterosaur: AMNH Library; Cuvier by Van Bree: De Agostini Picture Libra The Bridgeman Art Library. **Pages 12–13 A New Discovery.** Mary Anning by Donne: T Natural History Museum/The Image Works. **Pages 14–15 Fine Specimen.** *Pterodacty antiquus* fossil: AMNH FR 1942; Solnhofen quarry: AMNH; *Pterodactylus antiquus:* Peter Wellnhofer; Solnhofen lime quarry: Michael Nitzschke/AGE Fotostock; turtle fos Siepmann/AGE Fotostock. **Pages 16–17 Imperfect Evidence.** *Pterodactylus buffarinii* fos AMNH cast of Friuli Natural History Museum MFSN GP 1770; cockatoo: John Cancalosi/A Fotostock; sea gull: Arpad Radoczy/AGE Fotostock; wood thrush: BG Thomson/Scier Source; casting photo: AMNH/Rod Mickens. **Pages 20–21 The First Flying Reptiles.** Fr Arms Into Wings: Peters and Gutmann. **Pages 24–25 Inside the Nest.** *Pterodactylus eleg* fossil cast: AMNH FR 5147 pterosaur egg fossil cast: Institute of Vertebrate Paleontolc and Paleoanthropology, Chinese Academy of Sciences IVPP VI3758; corn snake egg a black swan egg: Dorling Kindersley/AGE Fotostock. **Pages 26–27 Locomotion on La** Pterosaur and dinosaur tracks cast: Polish Geological Institute; bat climbing: David Co Alamy; sloth: Juniors Bildarchiv/AGE Fotostock; roadrunner: Anthony Mercieca/Scier Source. **Pages 28–29 Into the Air.** Model Flyer: Archiv der Max-Planck-Gesellsch: Berlin-Dahlem; *Rhamphorhynchus Muensteri* fossil: AMNH FR 2323. **Pages 30–31 Wi Bones.** Flying squirrel: McDonald Wildlife/Animals Animals/AGE Fotostock; flying fi Tom Stack/WaterFrame/AGE Fotostock; flying lizard: Mint Frans Lanting/AGE Fotosto flying lemur: H. Brehm/Picture Alliance/Photoshot; flying frog: Photoshot/AGE Fotosto Three Kinds of Wings: Mick Ellison. **Pages 32–33 Built to Fly.** Wing bone/airplane wir © Julian Mulock. **Pages 34–35 Inside the Wings.** Pterosaur fossil photograph: Courte of Helmut Tischlinger; macaw feather: Kevin Schafer/NHPA/Photoshot; bat wing: C. Bra Picture Alliance/Photoshot. **Pages 36–37 Warm and Fuzzy.** Blue tit: Dave Bevan/Gard World Images/AGE Fotostock; *Sordes* hair courtesy of David M. Unwin. **Pages 38–Dragon of the Dawn.** Monument rocks: Corbis/AGE Fotostock, *Quetzalcoatlus northro* © John Sibbick. **Pages 40–41 Long Tall Pterosaur.** Douglas Lawson: From the Collectic of the Vertebrate Paleontology Laboratory, The University of Texas at Austin; Big Be National Park: Blomberg/Zoonar/AGE Fotostock. **Pages 42–43 Many Kinds of Cres** *Thalassodromeus sethi* fossil: cast of Earth Science Museum, Rio de Janeiro, DGM 147 R; *Germanodactylus* illustration: Dmitry Bogdanov/Wikimedia Commons. **Pages 44– What's the Point?** Peacock: iStockphoto; peacocking: Imagesource/AGE Fotostock. **Pag 46–47 Cooling and Steering.** *Thalassodromeus sethi* fossil: Alexander Kellner; sweati man: Classicstock/The Image Works; plane tail: Shutterstock. **Pages 48–49 Keel Ja** *Tropeognathus mesembrinus* fossil: reconstructed cast of National Museum of Brazil 6594 **Pages 50–51 Follow the Teeth.** Hornbill: Zoonar K. Bain/AGE Fotostock; *Dsungaripte* fossil: Alexander Kellner. **Pages 52–53 Filter Face.** Baleen whale: Christopher Swann/Scien Source; *Pterodaustro* fossil: Bayerische Staatssammlung für Paläntologie und Geolog photo by Frank Höck. **Pages 54–55 Creature Catcher.** Pteranodon fossil: AMNH FR 50! bat chasing insect: Stephen Dalton/NHPA/Photoshot; wolffish: R. Koenig/Blickwinkel/A(Fotostock; flamingo: C. Hütter/Picture Alliance/Photoshot; pelican: Mark Miller Photc AGE Fotostock; parrot: Sohns Imagebroker/AGE Fotostock. **Pages 56–57 Pterosaurs a Water.** Catching Fish: Alexander Kellner; How Did Pterosaurs Swim?: David W. E. Ho and Donald M. Henderson; Why Water?: Michael Black/Robert Harding Picture Library/A(Fotostock. **Pages 58–59 A Superb Site.** *Anhanguera santanae* fossil: John Conway; Upp Layer: Nodules (top and bottom): Alexander Kellner; Lower Layer: Slabs (top): Alexanc Kellner; Lower Layer: Slabs (bottom): John G. Maisey/AMNH. **Pages 60–61 How Do Foss Form?.** *Nemicolopterus crypticus* fossil: Cast of Institute for Vertebrate Paleontology a Paleoanthropology, Chinese Academy of Sciences IVPP V-I4377; Body Fossils: Mick Elliso AMNH; Trace Fossils: Mrs. Wang Shenna; External Molds: Sam Noble Museum/University Oklahoma; Internal Molds: Sam Noble Museum/University of Oklahoma. **Pages 62–63 T Next Big Thing?** Romanian fossils: Mick Ellison/AMNH; *Hatzegopteryx thambema* drawin Mark Witton; Pterosaur discovery site: Mark A. Norell/AMNH.

AMERICAN MUSEUM ᴼ NATURAL HISTORY

The American Museum of Natural History is one of the world's preeminent scientific, educational, and cultural institutions. Since the Museum was founded in 1869, collections have grown to include approximately 33 million specimens and artifacts. With 45 permanent exhibition halls, including the Rose Center for Earth and Space, t Hayden Planetarium, and the iconic Bernard Family Hall of North American Mammals, there is plenty to experience. The Museum also produces temporary exhibitions, rangi from *Extreme Mammals* to *Einstein* and *Darwin* to *Creatures of Light*, with recent exhibitions on poison and pterosaurs. The American Museum of Natural History is the o Ph.D. degree-granting museum in the Western Hemisphere, conferring a doctorate in Comparative Biology. **Visit amnh.org for more information.**